CONTRACTOR'S INSTALLATION LOG BOOK

Job Record Book

Steyerhaus Publications - Little Rock

STEYERHAUS
Publications

Introduction

Keeping track of completed jobs doesn't seem that important—until a customer calls about theirs weeks, months, or even years later. Detailed installation records are almost always a challenge. A customer asking for additional work done in the same finishes, colors, or style, or for repairs or maintenance of an installation done long ago can seem an almost impossible task without good installation records. Many times, contractors are forced to either rely on the information the customer provides (and trust that the information is correct), or go to the site to review your work and hope you can figure it out. The end result is never ideal.

This installation log book is the solution. Easily keep a record of customer information, installation details, materials used, colors and/or finishes, and other notes - all in one convenient, portable log book. With an easy-to-complete index included, you'll have all of your customer's information at your fingertips.

As a licensed contractor in Tucson, Arizona for many years, our business provided installation invoices to customers and kept a copy of them—but we STILL struggled with digging through them to find specific invoices, especially ones that weren't recent. And, even when we did find them, they might not have the information needed, such as the materials used on the job, the colors/finishes installed and specific requests from the customer. Nothing was more frustrating than having a customer who needed repair work done, new work added, or maintenance work done to their installation and not being able to find their information!

This installation log book was developed to address these exact problems. With this book, you'll be able to quickly find customer contact information, the start and completion dates, job price including the deposit received and balance due, a description of the job, what materials were installed, what colors and/or finishes were used, and any notes you made. You'll look professional and competent when you're able to quickly access their job information and provide details - even when the installation was completed months—or years—ago.

Each installation page in this log book includes a place for:

- Start & Completion Date
- Customer Information
- Price & Deposit Amount
- Balance Due
- Payment Date
- Installation Description
- Materials Installed
- Colors/Finishes
- Notes

In addition, every installation page has an accompanying dot grid sheet for drawings, calculations and additional notes. You'll also find an alphabetized index at the end to input your customers by name and page for easy reference and to be able to access your customers' job information quickly.

With a total of 100 job installation sheets, 100 dot grid sheets on the opposite pages for drawings and calculations, and a quick-reference index for easily accessing your customers' job information, this log book will help keep you organized, save time, and stay on the ball for your customers. Don't wait—start using this installation log book today and take control of your jobs!

Be sure to check out our companion book, Contractor's Estimate Log Book, to keep easy-to-access records of estimate details, including estimate amount, materials quoted, colors/finishes, dates and more!

Contractor's

Installation Log

Date Range

FROM: ______________________________

TO: ________________________________

Installation

Start Date: ______________________ Completion Date: ______________________

Customer Name: ____

Address: ____

Phone: _______________ Email: ____

Price: _________________________ Deposit Amount: _____________________

Balance: ________________________Date Paid: __________________________

Hours Worked: ____

Description of Installation: ____

____

____

____

Materials Installed: ____

____

____

____

____

Colors/Finishes: ____

____

____

____

Notes: ____

____

____

____

Installation

Start Date: ______________________ Completion Date: ____________________

Customer Name: ____

Address: ____

Phone: ______________ Email: ____

Price: __________________________ Deposit Amount: ____________________

Balance: _________________________Date Paid: ________________________

Hours Worked: ____

Description of Installation: ____

____

____

____

Materials Installed: ____

____

____

____

____

Colors/Finishes: ____

____

____

____

Notes: ____

____

____

____

Installation

Start Date: ____________________ Completion Date: ____________________

Customer Name: ____

Address: ____

Phone: ______________ Email: _____________________________________

Price: ________________________ Deposit Amount: ____________________

Balance: ______________________ Date Paid: _________________________

Hours Worked: ____

Description of Installation: _____________________________________

____

____

____

Materials Installed: ____

____

____

____

____

Colors/Finishes: ____

____

____

____

Notes: ____

____

____

____

Installation

Start Date: ______________________ Completion Date: ____________________

Customer Name: ____

Address: ____

Phone: ______________ Email: ____

Price: __________________________ Deposit Amount: ____________________

Balance: _________________________ Date Paid: _________________________

Hours Worked: ____

Description of Installation: ____

____

____

____

Materials Installed: ____

____

____

____

____

Colors/Finishes: ____

____

____

____

Notes: ____

____

____

____

Installation

Start Date: ______________________ Completion Date: ____________________

Customer Name: ____

Address: ____

Phone: ______________ Email: ____

Price: __________________________ Deposit Amount: _____________________

Balance: _________________________Date Paid: __________________________

Hours Worked: ____

Description of Installation: ____

____

____

____

Materials Installed: ____

____

____

____

____

Colors/Finishes: ____

____

____

____

Notes: ____

____

____

____

Installation

Start Date: ____________________ Completion Date: ____________________

Customer Name: ____

Address: ____

Phone: ______________ Email: ______________________________

Price: ______________________ Deposit Amount: ____________________

Balance: ______________________ Date Paid: ______________________

Hours Worked: ____

Description of Installation: ______________________________

____

____

____

Materials Installed: ____

____

____

____

____

Colors/Finishes: ____

____

____

____

Notes: ____

____

____

____

Installation

Start Date: ______________________ Completion Date: ______________________

Customer Name: ____

Address: ____

Phone: ______________ Email: ______________________________

Price: ______________________ Deposit Amount: ______________________

Balance: ______________________ Date Paid: ______________________

Hours Worked: ____

Description of Installation: ____

____

____

____

Materials Installed: ____

____

____

____

____

Colors/Finishes: ____

____

____

____

Notes: ____

____

____

____

Installation

Start Date: ______________________ Completion Date: ____________________

Customer Name: ____

Address: ____

Phone: ______________ Email: ____

Price: __________________________ Deposit Amount: ____________________

Balance: _________________________Date Paid: __________________________

Hours Worked: ____

Description of Installation: ______________________________________

____

____

____

Materials Installed: ____

____

____

____

____

Colors/Finishes: ____

____

____

____

Notes: ____

____

____

____

Installation

Start Date: ______________________ Completion Date: ______________________

Customer Name: ____

Address: ____

Phone: ______________ Email: ______________________________

Price: ______________________ Deposit Amount: ______________________

Balance: ______________________Date Paid: ______________________

Hours Worked: ____

Description of Installation: ____

____

____

____

Materials Installed: ____

____

____

____

____

Colors/Finishes: ____

____

____

____

Notes: ____

____

____

____

Installation

Start Date: ____________________ Completion Date: ____________________

Customer Name: ____

Address: ____

Phone: ______________ Email: ______________________________

Price: ____________________ Deposit Amount: ____________________

Balance: ____________________ Date Paid: ____________________

Hours Worked: ____

Description of Installation: ______________________________

____

____

____

Materials Installed: ______________________________

____

____

____

____

Colors/Finishes: ______________________________

____

____

____

Notes: ______________________________

____

____

____

Installation

Start Date: ______________________ Completion Date: ____________________

Customer Name: ____

Address: ____

Phone: ______________ Email: ____

Price: ___________________________ Deposit Amount: ____________________

Balance: ________________________ Date Paid: _________________________

Hours Worked: ____

Description of Installation: ____

____

____

____

Materials Installed: ____

____

____

____

____

Colors/Finishes: ____

____

____

____

Notes: ____

____

____

____

Installation

Start Date: ____________________ Completion Date: ____________________

Customer Name: ____

Address: ____

Phone: ______________ Email: ________________________________

Price: ____________________ Deposit Amount: ____________________

Balance: ____________________ Date Paid: ____________________

Hours Worked: ____

Description of Installation: ________________________________

____

____

____

Materials Installed: ________________________________

____

____

____

____

Colors/Finishes: ________________________________

____

____

____

Notes: ____

____

____

____

Installation

Start Date: ______________________ Completion Date: ______________________

Customer Name: ____

Address: ____

Phone: ______________ Email: ______________________________________

Price: ________________________ Deposit Amount: ______________________

Balance: ________________________ Date Paid: ______________________

Hours Worked: ____

Description of Installation: ______________________________________

____

____

____

Materials Installed: ____

____

____

____

____

Colors/Finishes: ____

____

____

____

Notes: ____

____

____

____

Installation

Start Date: ______________________ Completion Date: ______________________

Customer Name: ____

Address: ____

Phone: ______________ Email: ____

Price: __________________________ Deposit Amount: ______________________

Balance: _________________________ Date Paid: ___________________________

Hours Worked: ____

Description of Installation: ____

____

____

____

Materials Installed: ____

____

____

____

____

Colors/Finishes: ____

____

____

____

Notes: ____

____

____

____

Installation

Start Date: ______________________ Completion Date: ____________________

Customer Name: ____

Address: ____

Phone: _______________ Email: _______________________________

Price: __________________________ Deposit Amount: _____________________

Balance: _________________________ Date Paid: __________________________

Hours Worked: ____

Description of Installation: ______________________________________

____

____

____

Materials Installed: ____

____

____

____

____

Colors/Finishes: ____

____

____

____

Notes: ____

____

____

____

Installation

Start Date: ______________________ Completion Date: ____________________

Customer Name: ____

Address: ____

Phone: ______________ Email: _______________________________________

Price: __________________________ Deposit Amount: ____________________

Balance: _________________________Date Paid: _________________________

Hours Worked: ____

Description of Installation: ____

____

____

____

Materials Installed: ____

____

____

____

____

Colors/Finishes: ____

____

____

____

Notes: ____

____

____

____

Installation

Start Date: ______________________ Completion Date: ______________________

Customer Name: ____

Address: ____

Phone: _______________ Email: ______________________________

Price: ___________________________ Deposit Amount: ______________________

Balance: _________________________Date Paid: ____________________________

Hours Worked: ____

Description of Installation: ____

____

____

____

Materials Installed: ____

____

____

____

____

Colors/Finishes: ____

____

____

____

Notes: ____

____

____

____

Installation

Start Date: ______________________ Completion Date: ____________________

Customer Name: ____

Address: ____

Phone: ______________ Email: _____________________________________

Price: ___________________________ Deposit Amount: ____________________

Balance: _________________________ Date Paid: ________________________

Hours Worked: ____

Description of Installation: _______________________________________

____

____

____

Materials Installed: ____

____

____

____

____

Colors/Finishes: ____

____

____

____

Notes: ____

____

____

____

Installation

Start Date: ______________________ Completion Date: ____________________

Customer Name: ____

Address: ____

Phone: _______________ Email: _________________________________

Price: __________________________ Deposit Amount: ____________________

Balance: _________________________Date Paid: __________________________

Hours Worked: ____

Description of Installation: ______________________________________

____

____

____

Materials Installed: ____

____

____

____

____

Colors/Finishes: ____

____

____

____

Notes: ____

____

____

____

Installation

Start Date: ______________________ Completion Date: ____________________

Customer Name: ____

Address: ____

Phone: ______________ Email: _______________________________________

Price: _________________________ Deposit Amount: _____________________

Balance: _______________________Date Paid: __________________________

Hours Worked: ____

Description of Installation: ____

____

____

____

Materials Installed: ____

____

____

____

____

Colors/Finishes: ____

____

____

____

Notes: ____

____

____

____

Installation

Start Date: ______________________ Completion Date: ______________________

Customer Name: ____

Address: ____

Phone: ______________ Email: ______________________________

Price: ______________________ Deposit Amount: ______________________

Balance: ______________________ Date Paid: ______________________

Hours Worked: ____

Description of Installation: ____

____

____

____

Materials Installed: ____

____

____

____

____

Colors/Finishes: ____

____

____

____

Notes: ____

____

____

____

Installation

Start Date: ______________________ Completion Date: ______________________

Customer Name: ____

Address: ____

Phone: ______________ Email: _____________________________________

Price: __________________________ Deposit Amount: ______________________

Balance: ________________________ Date Paid: ___________________________

Hours Worked: ____

Description of Installation: ____

____

____

____

Materials Installed: ____

____

____

____

____

Colors/Finishes: ____

____

____

____

Notes: ____

____

____

____

Installation

Start Date: ______________________ Completion Date: ______________________

Customer Name: ____

Address: ____

Phone: ______________ Email: ____

Price: _________________________ Deposit Amount: ______________________

Balance: ________________________Date Paid: ___________________________

Hours Worked: ____

Description of Installation: ____

____

____

____

Materials Installed: ____

____

____

____

____

Colors/Finishes: ____

____

____

____

Notes: ____

____

____

____

Installation

Start Date: ______________________ Completion Date: ____________________

Customer Name: ____

Address: ____

Phone: ______________ Email: ______________________________________

Price: __________________________ Deposit Amount: ____________________

Balance: ________________________Date Paid: ________________________

Hours Worked: ____

Description of Installation: ____

____

____

____

Materials Installed: ____

____

____

____

____

Colors/Finishes: ____

____

____

____

Notes: ____

____

____

____

Installation

Start Date: ______________________ Completion Date: ____________________

Customer Name: ____

Address: ____

Phone: ______________ Email: ______________________________________

Price: _________________________ Deposit Amount: ____________________

Balance: ________________________Date Paid: __________________________

Hours Worked: ____

Description of Installation: ____________________________________

____

____

____

Materials Installed: ____

____

____

____

____

Colors/Finishes: ____

____

____

____

Notes: ____

____

____

____

Installation

Start Date: ______________________ Completion Date: ______________________

Customer Name: ____

Address: ____

Phone: _______________ Email: ____

Price: __________________________ Deposit Amount: ______________________

Balance: _________________________Date Paid: ___________________________

Hours Worked: ____

Description of Installation: ____

____

____

____

Materials Installed: ____

____

____

____

____

Colors/Finishes: ____

____

____

____

Notes: ____

____

____

____

Installation

Start Date: ______________________ Completion Date: _____________________

Customer Name: ____

Address: ____

Phone: _______________ Email: ____

Price: __________________________ Deposit Amount: ____________________

Balance: ________________________Date Paid: ___________________________

Hours Worked: ____

Description of Installation: ____

____

____

____

Materials Installed: ____

____

____

____

____

Colors/Finishes: ____

____

____

____

Notes: ____

____

____

____

Installation

Start Date: ______________________ Completion Date: ______________________

Customer Name: ____

Address: ____

Phone: _______________ Email: _____________________________________

Price: __________________________ Deposit Amount: ______________________

Balance: ________________________ Date Paid: ____________________________

Hours Worked: ____

Description of Installation: ____

____

____

____

Materials Installed: ____

____

____

____

____

Colors/Finishes: ____

____

____

____

Notes: ____

____

____

____

Installation

Start Date: ______________________ Completion Date: ______________________

Customer Name: ____

Address: ____

Phone: _______________ Email: ________________________________

Price: ________________________ Deposit Amount: ______________________

Balance: ________________________ Date Paid: __________________________

Hours Worked: ____

Description of Installation: ___________________________________

____

____

____

Materials Installed: ____

____

____

____

____

Colors/Finishes: ____

____

____

____

Notes: ____

____

____

____

Installation

Start Date: ______________________ Completion Date: ____________________

Customer Name: ____

Address: ____

Phone: _______________ Email: ______________________________________

Price: __________________________ Deposit Amount: _____________________

Balance: ________________________Date Paid: __________________________

Hours Worked: ____

Description of Installation: ____

____

____

____

Materials Installed: ____

____

____

____

____

Colors/Finishes: ____

____

____

____

Notes: ____

____

____

____

Installation

Start Date: ______________________ Completion Date: ____________________

Customer Name: ____

Address: ____

Phone: ______________ Email: ____

Price: __________________________ Deposit Amount: _____________________

Balance: ________________________ Date Paid: __________________________

Hours Worked: ____

Description of Installation: ____

____

____

____

Materials Installed: ____

____

____

____

____

Colors/Finishes: ____

____

____

____

Notes: ____

____

____

____

Installation

Start Date: ______________________ Completion Date: ______________________

Customer Name: ____

Address: ____

Phone: ______________ Email: ____

Price: __________________________ Deposit Amount: _____________________

Balance: ________________________ Date Paid: __________________________

Hours Worked: ____

Description of Installation: ____

____

____

____

Materials Installed: ____

____

____

____

____

Colors/Finishes: ____

____

____

____

Notes: ____

____

____

____

Installation

Start Date: ______________________ Completion Date: ______________________

Customer Name: ____

Address: ____

Phone: _______________ Email: ____

Price: ___________________________ Deposit Amount: ______________________

Balance: _________________________ Date Paid: ___________________________

Hours Worked: ____

Description of Installation: ____

____

____

____

Materials Installed: ____

____

____

____

____

Colors/Finishes: ____

____

____

____

Notes: ____

____

____

____

Installation

Start Date: ______________________ Completion Date: ____________________

Customer Name: ____

Address: ____

Phone: ______________ Email: ____

Price: __________________________ Deposit Amount: ___________________

Balance: ________________________ Date Paid: _________________________

Hours Worked: ____

Description of Installation: ____

____

____

____

Materials Installed: ____

____

____

____

____

Colors/Finishes: ____

____

____

____

Notes: ____

____

____

____

Installation

Start Date: ______________________ Completion Date: ____________________

Customer Name: ____

Address: ____

Phone: ______________ Email: ____

Price: __________________________ Deposit Amount: _____________________

Balance: ________________________Date Paid: __________________________

Hours Worked: ____

Description of Installation: ____

____

____

____

Materials Installed: ____

____

____

____

____

Colors/Finishes: ____

____

____

____

Notes: ____

____

____

____

Installation

Start Date: ______________________ Completion Date: ______________________

Customer Name: ____

Address: ____

Phone: _______________ Email: ____

Price: ___________________________ Deposit Amount: ______________________

Balance: _________________________ Date Paid: ___________________________

Hours Worked: ____

Description of Installation: ____

____

____

____

Materials Installed: ____

____

____

____

____

Colors/Finishes: ____

____

____

____

Notes: ____

____

____

____

Installation

Start Date: ______________________ Completion Date: ______________________

Customer Name: ____

Address: ____

Phone: _______________ Email: ____

Price: ___________________________ Deposit Amount: ______________________

Balance: _________________________ Date Paid: ___________________________

Hours Worked: ____

Description of Installation: ____

____

____

____

Materials Installed: ____

____

____

____

____

Colors/Finishes: ____

____

____

____

Notes: ____

____

____

____

Installation

Start Date: ______________________ Completion Date: ____________________

Customer Name: ____

Address: ____

Phone: ______________ Email: ___________________________________

Price: __________________________ Deposit Amount: _____________________

Balance: ________________________Date Paid: __________________________

Hours Worked: ____

Description of Installation: _____________________________________

____

____

____

Materials Installed: ____

____

____

____

____

Colors/Finishes: ____

____

____

____

Notes: ____

____

____

____

Installation

Start Date: ______________________ Completion Date: ______________________

Customer Name: ____

Address: ____

Phone: _______________ Email: ____

Price: ___________________________ Deposit Amount: _____________________

Balance: _________________________ Date Paid: ___________________________

Hours Worked: ____

Description of Installation: ____

____

____

____

Materials Installed: ____

____

____

____

____

Colors/Finishes: ____

____

____

____

Notes: ____

____

____

____

Installation

Start Date: ______________________ Completion Date: ______________________

Customer Name: ____

Address: ____

Phone: ______________ Email: ______________________________

Price: ______________________ Deposit Amount: ______________________

Balance: ______________________ Date Paid: ______________________

Hours Worked: ____

Description of Installation: ____

____

____

____

Materials Installed: ____

____

____

____

____

Colors/Finishes: ____

____

____

____

Notes: ____

____

____

____

Installation

Start Date: ______________________ Completion Date: ____________________

Customer Name: ____

Address: ____

Phone: ______________ Email: _______________________________________

Price: ___________________________ Deposit Amount: _____________________

Balance: __________________________ Date Paid: ___________________________

Hours Worked: ____

Description of Installation: ____

____

____

____

Materials Installed: ____

____

____

____

____

Colors/Finishes: ____

____

____

____

Notes: ____

____

____

____

Installation

Start Date: ______________________ Completion Date: ______________________

Customer Name: ____

Address: ____

Phone: _______________ Email: ____

Price: __________________________ Deposit Amount: ______________________

Balance: ________________________ Date Paid: ____________________________

Hours Worked: ____

Description of Installation: ____

____

____

____

Materials Installed: ____

____

____

____

____

Colors/Finishes: ____

____

____

____

Notes: ____

____

____

____

Installation

Start Date: ____________________ Completion Date: ____________________

Customer Name: ____

Address: ____

Phone: _____________ Email: ____

Price: ________________________ Deposit Amount: ____________________

Balance: ______________________ Date Paid: ________________________

Hours Worked: ____

Description of Installation: ____

____

____

____

Materials Installed: ____

____

____

____

____

Colors/Finishes: ____

____

____

____

Notes: ____

____

____

____

Installation

Start Date: ____________________ Completion Date: ____________________

Customer Name: ____

Address: ____

Phone: ______________ Email: ______________________________

Price: ____________________ Deposit Amount: ____________________

Balance: ____________________ Date Paid: ____________________

Hours Worked: ____

Description of Installation: ____

____

____

____

Materials Installed: ____

____

____

____

____

Colors/Finishes: ____

____

____

____

Notes: ____

____

____

____

Installation

Start Date: ______________________ Completion Date: ____________________

Customer Name: ____

Address: ____

Phone: _______________ Email: ____

Price: __________________________ Deposit Amount: _____________________

Balance: ________________________ Date Paid: _________________________

Hours Worked: ____

Description of Installation: ____

____

____

____

Materials Installed: ____

____

____

____

____

Colors/Finishes: ____

____

____

____

Notes: ____

____

____

____

Installation

Start Date: ______________________ Completion Date: ____________________

Customer Name: ____

Address: ____

Phone: ______________ Email: ______________________________________

Price: __________________________ Deposit Amount: ____________________

Balance: ________________________Date Paid: _________________________

Hours Worked: ____

Description of Installation: _______________________________________

____

____

____

Materials Installed: ____

____

____

____

____

Colors/Finishes: ____

____

____

____

Notes: ____

____

____

____

Installation

Start Date: ______________________ Completion Date: ____________________

Customer Name: ____

Address: ____

Phone: ______________ Email: ____

Price: __________________________ Deposit Amount: _____________________

Balance: ________________________Date Paid: ___________________________

Hours Worked: ____

Description of Installation: ____

____

____

____

Materials Installed: ____

____

____

____

____

Colors/Finishes: ____

____

____

____

Notes: ____

____

____

____

Installation

Start Date: ______________________ Completion Date: ____________________

Customer Name: ____

Address: ____

Phone: ______________ Email: ____

Price: ________________________ Deposit Amount: ____________________

Balance: _______________________Date Paid: _________________________

Hours Worked: ____

Description of Installation: ____

____

____

____

Materials Installed: ____

____

____

____

____

Colors/Finishes: ____

____

____

____

Notes: ____

____

____

____

Installation

Start Date: ______________________ Completion Date: ____________________

Customer Name: ____

Address: ____

Phone: _______________ Email: ____

Price: __________________________ Deposit Amount: _____________________

Balance: _________________________Date Paid: ___________________________

Hours Worked: ____

Description of Installation: ____

____

____

____

Materials Installed: ____

____

____

____

____

Colors/Finishes: ____

____

____

____

Notes: ____

____

____

____

Installation

Start Date: ______________________ Completion Date: ____________________

Customer Name: ____

Address: ____

Phone: ______________ Email: _______________________________________

Price: __________________________ Deposit Amount: _____________________

Balance: ________________________Date Paid: ___________________________

Hours Worked: ____

Description of Installation: ____

____

____

____

Materials Installed: ____

____

____

____

____

Colors/Finishes: ____

____

____

____

Notes: ____

____

____

____

Installation

Start Date: ______________________ Completion Date: _____________________

Customer Name: ____

Address: ____

Phone: _______________ Email: ____

Price: __________________________ Deposit Amount: _____________________

Balance: _________________________Date Paid: ___________________________

Hours Worked: ____

Description of Installation: ____

____

____

____

Materials Installed: ____

____

____

____

____

Colors/Finishes: ____

____

____

____

Notes: ____

____

____

____

Installation

Start Date: ______________________ Completion Date: ______________________

Customer Name: ____

Address: ____

Phone: ______________ Email: ______________________________________

Price: __________________________ Deposit Amount: ______________________

Balance: _________________________ Date Paid: __________________________

Hours Worked: ____

Description of Installation: ______________________________________

____

____

____

Materials Installed: ____

____

____

____

____

Colors/Finishes: ____

____

____

____

Notes: ____

____

____

____

Installation

Start Date: ______________________ Completion Date: ______________________

Customer Name: ____

Address: ____

Phone: _______________ Email: ____

Price: __________________________ Deposit Amount: ______________________

Balance: ________________________Date Paid: ___________________________

Hours Worked: ____

Description of Installation: ____

____

____

____

Materials Installed: ____

____

____

____

____

Colors/Finishes: ____

____

____

____

Notes: ____

____

____

____

Installation

Start Date: ______________________ Completion Date: ______________________

Customer Name: ____

Address: ____

Phone: _______________ Email: ____________________________________

Price: __________________________ Deposit Amount: ______________________

Balance: ________________________ Date Paid: ___________________________

Hours Worked: ____

Description of Installation: ____

____

____

____

Materials Installed: ____

____

____

____

____

Colors/Finishes: ____

____

____

____

Notes: ____

____

____

____

Installation

Start Date: ______________________ Completion Date: ______________________

Customer Name: ____

Address: ____

Phone: ______________ Email: ______________________________

Price: ______________________ Deposit Amount: ______________________

Balance: ______________________ Date Paid: ______________________

Hours Worked: ____

Description of Installation: ____

____

____

____

Materials Installed: ____

____

____

____

____

Colors/Finishes: ____

____

____

____

Notes: ____

____

____

____

Installation

Start Date: ______________________ Completion Date: ______________________

Customer Name: ____

Address: ____

Phone: _______________ Email: ____

Price: __________________________ Deposit Amount: _____________________

Balance: _________________________ Date Paid: ___________________________

Hours Worked: ____

Description of Installation: ____

____

____

____

Materials Installed: ____

____

____

____

____

Colors/Finishes: ____

____

____

____

Notes: ____

____

____

____

Installation

Start Date: ______________________ Completion Date: ______________________

Customer Name: ____

Address: ____

Phone: ______________ Email: ______________________________

Price: ______________________ Deposit Amount: ______________________

Balance: ______________________ Date Paid: ______________________

Hours Worked: ____

Description of Installation: ____

____

____

____

Materials Installed: ____

____

____

____

____

Colors/Finishes: ____

____

____

____

Notes: ____

____

____

____

Installation

Start Date: ______________________ Completion Date: ______________________

Customer Name: ____

Address: ____

Phone: _______________ Email: ____

Price: ___________________________ Deposit Amount: ____________________

Balance: __________________________Date Paid: _________________________

Hours Worked: ____

Description of Installation: ____

____

____

____

Materials Installed: ____

____

____

____

____

Colors/Finishes: ____

____

____

____

Notes: ____

____

____

____

Installation

Start Date: ______________________ Completion Date: _____________________

Customer Name: ____

Address: ____

Phone: ______________ Email: ____

Price: ___________________________ Deposit Amount: ____________________

Balance: __________________________Date Paid: __________________________

Hours Worked: ____

Description of Installation: ____

____

____

____

Materials Installed: ____

____

____

____

____

Colors/Finishes: ____

____

____

____

Notes: ____

____

____

____

Installation

Start Date: ______________________ Completion Date: ____________________

Customer Name: ____

Address: ____

Phone: ______________ Email: ________________________________

Price: ___________________________ Deposit Amount: _______________

Balance: _________________________Date Paid: ___________________

Hours Worked: ____

Description of Installation: ______________________________________

____

____

____

Materials Installed: ____

____

____

____

____

Colors/Finishes: ____

____

____

____

Notes: ____

____

____

____

Installation

Start Date: ______________________ Completion Date: _____________________

Customer Name: ____

Address: ____

Phone: _______________ Email: ____

Price: ___________________________ Deposit Amount: ____________________

Balance: _________________________ Date Paid: ___________________________

Hours Worked: ____

Description of Installation: ____

____

____

____

Materials Installed: ____

____

____

____

____

Colors/Finishes: ____

____

____

____

Notes: ____

____

____

____

Installation

Start Date: ______________________ Completion Date: ____________________

Customer Name: ____

Address: ____

Phone: ______________ Email: ______________________________________

Price: __________________________ Deposit Amount: ____________________

Balance: ________________________Date Paid: _________________________

Hours Worked: ____

Description of Installation: ____

____

____

____

Materials Installed: ____

____

____

____

____

Colors/Finishes: ____

____

____

____

Notes: ____

____

____

____

Installation

Start Date: ______________________ Completion Date: ______________________

Customer Name: ____

Address: ____

Phone: _______________ Email: ____

Price: ___________________________ Deposit Amount: ______________________

Balance: _________________________Date Paid: ____________________________

Hours Worked: ____

Description of Installation: ____

____

____

____

Materials Installed: ____

____

____

____

____

Colors/Finishes: ____

____

____

____

Notes: ____

____

____

____

Installation

Start Date: ____________________ Completion Date: ____________________

Customer Name: ____

Address: ____

Phone: ______________ Email: ____________________________

Price: ____________________ Deposit Amount: ____________________

Balance: ____________________ Date Paid: ____________________

Hours Worked: ____

Description of Installation: ____

____

____

____

Materials Installed: ____

____

____

____

____

Colors/Finishes: ____

____

____

____

Notes: ____

____

____

____

Installation

Start Date: ______________________ Completion Date: ______________________

Customer Name: ____

Address: ____

Phone: ______________ Email: ______________________________________

Price: ______________________ Deposit Amount: ______________________

Balance: ______________________ Date Paid: ______________________

Hours Worked: ____

Description of Installation: ____

____

____

____

Materials Installed: ____

____

____

____

____

Colors/Finishes: ____

____

____

____

Notes: ____

____

____

____

Installation

Start Date: ______________________ Completion Date: ____________________

Customer Name: ____

Address: ____

Phone: _____________ Email: ______________________________________

Price: _________________________ Deposit Amount: ____________________

Balance: ________________________Date Paid: ________________________

Hours Worked: ____

Description of Installation: ______________________________________

____

____

____

Materials Installed: ____

____

____

____

____

Colors/Finishes: ____

____

____

____

Notes: ____

____

____

____

Installation

Start Date: ______________________ Completion Date: ______________________

Customer Name: ____

Address: ____

Phone: _______________ Email: ____

Price: ___________________________ Deposit Amount: ____________________

Balance: _________________________ Date Paid: __________________________

Hours Worked: ____

Description of Installation: ____

____

____

____

Materials Installed: ____

____

____

____

____

Colors/Finishes: ____

____

____

____

Notes: ____

____

____

____

Installation

Start Date: ________________________ Completion Date: ______________________

Customer Name: ____

Address: ____

Phone: ________________ Email: ____

Price: ____________________________ Deposit Amount: ________________________

Balance: __________________________Date Paid: _____________________________

Hours Worked: ____

Description of Installation: ____

____

____

____

Materials Installed: ____

____

____

____

____

Colors/Finishes: ____

____

____

____

Notes: ____

____

____

____

Installation

Start Date: ______________________ Completion Date: ____________________

Customer Name: ____

Address: ____

Phone: ______________ Email: _______________________________________

Price: __________________________ Deposit Amount: _____________________

Balance: ________________________ Date Paid: _________________________

Hours Worked: ____

Description of Installation: _____________________________________

____

____

____

Materials Installed: ____

____

____

____

____

Colors/Finishes: ____

____

____

____

Notes: ____

____

____

____

Installation

Start Date: ______________________ Completion Date: ______________________

Customer Name: ____

Address: ____

Phone: _______________ Email: ____

Price: __________________________ Deposit Amount: ______________________

Balance: _________________________ Date Paid: ___________________________

Hours Worked: ____

Description of Installation: ____

____

____

____

Materials Installed: ____

____

____

____

____

Colors/Finishes: ____

____

____

____

Notes: ____

____

____

____

Installation

Start Date: ______________________ Completion Date: ______________________

Customer Name: ____

Address: ____

Phone: ______________ Email: ______________________________

Price: __________________________ Deposit Amount: ______________________

Balance: ________________________ Date Paid: ___________________________

Hours Worked: ____

Description of Installation: ______________________________________

____

____

____

Materials Installed: ____

____

____

____

____

Colors/Finishes: ____

____

____

____

Notes: ____

____

____

____

Installation

Start Date: ______________________ Completion Date: ______________________

Customer Name: ____

Address: ____

Phone: ______________ Email: ____

Price: ___________________________ Deposit Amount: ____________________

Balance: _________________________ Date Paid: __________________________

Hours Worked: ____

Description of Installation: ____

____

____

____

Materials Installed: ____

____

____

____

____

Colors/Finishes: ____

____

____

____

Notes: ____

____

____

____

Installation

Start Date: ______________________ Completion Date: ____________________

Customer Name: ____

Address: ____

Phone: ______________ Email: ______________________________________

Price: __________________________ Deposit Amount: ____________________

Balance: ________________________ Date Paid: ________________________

Hours Worked: ____

Description of Installation: ____

____

____

____

Materials Installed: ____

____

____

____

____

Colors/Finishes: ____

____

____

____

Notes: ____

____

____

____

Installation

Start Date: ______________________ Completion Date: ______________________

Customer Name: ____

Address: ____

Phone: ______________ Email: ______________________________

Price: ______________________ Deposit Amount: ______________________

Balance: ______________________ Date Paid: ______________________

Hours Worked: ____

Description of Installation: ____

____

____

____

Materials Installed: ____

____

____

____

____

Colors/Finishes: ____

____

____

____

Notes: ____

____

____

____

Installation

Start Date: ______________________ Completion Date: _____________________

Customer Name: ____

Address: ____

Phone: _______________ Email: _____________________________________

Price: __________________________ Deposit Amount: ___________________

Balance: ________________________ Date Paid: ________________________

Hours Worked: ____

Description of Installation: ____

____

____

____

Materials Installed: ____

____

____

____

____

Colors/Finishes: ____

____

____

____

Notes: ____

____

____

____

Installation

Start Date: ______________________ Completion Date: ____________________

Customer Name: ____

Address: ____

Phone: _______________ Email: ____

Price: ___________________________ Deposit Amount: ___________________

Balance: _________________________ Date Paid: __________________________

Hours Worked: ____

Description of Installation: ____

____

____

____

Materials Installed: ____

____

____

____

____

Colors/Finishes: ____

____

____

____

Notes: ____

____

____

____

Installation

Start Date: ______________________ Completion Date: ______________________

Customer Name: ____

Address: ____

Phone: _______________ Email: ____

Price: __________________________ Deposit Amount: ______________________

Balance: ________________________ Date Paid: ___________________________

Hours Worked: ____

Description of Installation: ____

____

____

____

Materials Installed: ____

____

____

____

____

Colors/Finishes: ____

____

____

____

Notes: ____

____

____

____

Installation

Start Date: ______________________ Completion Date: ____________________

Customer Name: ____

Address: ____

Phone: _______________ Email: ____________________________________

Price: __________________________ Deposit Amount: ____________________

Balance: ________________________Date Paid: __________________________

Hours Worked: ____

Description of Installation: ______________________________________

____

____

____

Materials Installed: ____

____

____

____

____

Colors/Finishes: ____

____

____

____

Notes: ____

____

____

____

Installation

Start Date: ______________________ Completion Date: ____________________

Customer Name: ____

Address: ____

Phone: ______________ Email: ____

Price: __________________________ Deposit Amount: _____________________

Balance: __________________________Date Paid: _________________________

Hours Worked: ____

Description of Installation: ____

____

____

____

Materials Installed: ____

____

____

____

____

Colors/Finishes: ____

____

____

____

Notes: ____

____

____

____

Installation

Start Date: ______________________ Completion Date: _____________________

Customer Name: ____

Address: ____

Phone: _______________ Email: ____

Price: ___________________________ Deposit Amount: ____________________

Balance: _________________________ Date Paid: ___________________________

Hours Worked: ____

Description of Installation: ____

____

____

____

Materials Installed: ____

____

____

____

____

Colors/Finishes: ____

____

____

____

Notes: ____

____

____

____

Installation

Start Date: ______________________ Completion Date: ______________________

Customer Name: ____

Address: ____

Phone: _______________ Email: ____

Price: ___________________________ Deposit Amount: ______________________

Balance: _________________________Date Paid: ___________________________

Hours Worked: ____

Description of Installation: ____

____

____

____

Materials Installed: ____

____

____

____

____

Colors/Finishes: ____

____

____

____

Notes: ____

____

____

____

Installation

Start Date: ______________________ Completion Date: ____________________

Customer Name: ____

Address: ____

Phone: ______________ Email: ________________________________

Price: __________________________ Deposit Amount: _____________________

Balance: ________________________ Date Paid: _______________________

Hours Worked: ____

Description of Installation: _____________________________________

____

____

____

Materials Installed: ____

____

____

____

____

Colors/Finishes: ____

____

____

____

Notes: ____

____

____

____

Installation

Start Date: ______________________ Completion Date: ______________________

Customer Name: ____

Address: ____

Phone: _______________ Email: _______________________________________

Price: ___________________________ Deposit Amount: ____________________

Balance: __________________________Date Paid: __________________________

Hours Worked: ____

Description of Installation: ____

____

____

____

Materials Installed: ____

____

____

____

____

Colors/Finishes: ____

____

____

____

Notes: ____

____

____

____

Installation

Start Date: ______________________ Completion Date: ____________________

Customer Name: ____

Address: ____

Phone: ______________ Email: ____

Price: __________________________ Deposit Amount: ____________________

Balance: _________________________Date Paid: _________________________

Hours Worked: ____

Description of Installation: ____

____

____

____

Materials Installed: ____

____

____

____

____

Colors/Finishes: ____

____

____

____

Notes: ____

____

____

____

Installation

Start Date: ______________________ Completion Date: ____________________

Customer Name: ____

Address: ____

Phone: _______________ Email: ____

Price: ___________________________ Deposit Amount: ____________________

Balance: __________________________Date Paid: _________________________

Hours Worked: ____

Description of Installation: ____

____

____

____

Materials Installed: ____

____

____

____

____

Colors/Finishes: ____

____

____

____

Notes: ____

____

____

____

Installation

Start Date: ________________________ Completion Date: ______________________

Customer Name: ____

Address: ____

Phone: ________________ Email: ____

Price: ____________________________ Deposit Amount: _______________________

Balance: ___________________________ Date Paid: _____________________________

Hours Worked: ____

Description of Installation: ____

Materials Installed: ____

Colors/Finishes: ____

Notes: ____

Installation

Start Date: ______________________ Completion Date: _____________________

Customer Name: ____

Address: ____

Phone: _______________ Email: ______________________________________

Price: __________________________ Deposit Amount: ______________________

Balance: ________________________ Date Paid: ___________________________

Hours Worked: ____

Description of Installation: ____

____

____

____

Materials Installed: ____

____

____

____

____

Colors/Finishes: ____

____

____

____

Notes: ____

____

____

____

Installation

Start Date: ______________________ Completion Date: ______________________

Customer Name: ____

Address: ____

Phone: _______________ Email: ____

Price: ___________________________ Deposit Amount: ___________________

Balance: __________________________ Date Paid: _______________________

Hours Worked: ____

Description of Installation: ____

____

____

____

Materials Installed: ____

____

____

____

____

Colors/Finishes: ____

____

____

____

Notes: ____

____

____

____

Installation

Start Date: ______________________ Completion Date: ______________________

Customer Name: ____

Address: ____

Phone: _______________ Email: ____

Price: __________________________ Deposit Amount: ______________________

Balance: _________________________Date Paid: __________________________

Hours Worked: ____

Description of Installation: ____

____

____

____

Materials Installed: ____

____

____

____

____

Colors/Finishes: ____

____

____

____

Notes: ____

____

____

____

Installation

Start Date: ______________________ Completion Date: ____________________

Customer Name: ____

Address: ____

Phone: ______________ Email: ____

Price: __________________________ Deposit Amount: _____________________

Balance: ________________________Date Paid: _________________________

Hours Worked: ____

Description of Installation: ____

____

____

____

Materials Installed: ____

____

____

____

____

Colors/Finishes: ____

____

____

____

Notes: ____

____

____

____

Installation

Start Date: ______________________ Completion Date: ____________________

Customer Name: ____

Address: ____

Phone: ______________ Email: ______________________________________

Price: __________________________ Deposit Amount: ____________________

Balance: ________________________ Date Paid: ________________________

Hours Worked: ____

Description of Installation: ______________________________________

____

____

____

Materials Installed: ____

____

____

____

____

Colors/Finishes: ____

____

____

____

Notes: ____

____

____

____

Installation

Start Date: ______________________ Completion Date: ____________________

Customer Name: ____

Address: ____

Phone: ______________ Email: _______________________________________

Price: __________________________ Deposit Amount: ____________________

Balance: ________________________ Date Paid: _________________________

Hours Worked: ____

Description of Installation: ____

____

____

____

Materials Installed: ____

____

____

____

____

Colors/Finishes: ____

____

____

____

Notes: ____

____

____

____

Installation

Start Date: ______________________ Completion Date: ____________________

Customer Name: ____

Address: ____

Phone: ______________ Email: _________________________________

Price: __________________________ Deposit Amount: ___________________

Balance: ________________________Date Paid: _______________________

Hours Worked: ____

Description of Installation: ____________________________________

____

____

____

Materials Installed: ____

____

____

____

____

Colors/Finishes: ____

____

____

____

Notes: ____

____

____

____

Installation

Start Date: ______________________ Completion Date: ____________________

Customer Name: ____

Address: ____

Phone: ______________ Email: ________________________________

Price: _________________________ Deposit Amount: ___________________

Balance: ________________________Date Paid: _______________________

Hours Worked: ____

Description of Installation: ____________________________________

____

____

____

Materials Installed: ____

____

____

____

____

Colors/Finishes: ____

____

____

____

Notes: ____

____

____

____

Installation

Start Date: ______________________ Completion Date: ______________________

Customer Name: ____

Address: ____

Phone: _______________ Email: ____

Price: ___________________________ Deposit Amount: ______________________

Balance: __________________________Date Paid: ___________________________

Hours Worked: ____

Description of Installation: ____

____

____

____

Materials Installed: ____

____

____

____

____

Colors/Finishes: ____

____

____

____

Notes: ____

____

____

____

Installation

Start Date: ______________________ Completion Date: ______________________

Customer Name: ____

Address: ____

Phone: ______________ Email: ______________________________________

Price: ______________________ Deposit Amount: ______________________

Balance: ______________________ Date Paid: ______________________

Hours Worked: ____

Description of Installation: ______________________________________

____

____

____

Materials Installed: ____

____

____

____

____

Colors/Finishes: ____

____

____

____

Notes: ____

____

____

____

Installation

Start Date: ______________________ Completion Date: _____________________

Customer Name: ____

Address: ____

Phone: _______________ Email: ____

Price: __________________________ Deposit Amount: _____________________

Balance: _________________________ Date Paid: _________________________

Hours Worked: ____

Description of Installation: ____

____

____

____

Materials Installed: ____

____

____

____

____

Colors/Finishes: ____

____

____

____

Notes: ____

____

____

____

Installation

Start Date: ______________________ Completion Date: ______________________

Customer Name: ____

Address: ____

Phone: ______________ Email: ________________________________

Price: __________________________ Deposit Amount: ______________________

Balance: ________________________Date Paid: ________________________

Hours Worked: ____

Description of Installation: ____

____

____

____

Materials Installed: ____

____

____

____

____

Colors/Finishes: ____

____

____

____

Notes: ____

____

____

____

Installation

Start Date: ______________________ Completion Date: ______________________

Customer Name: ____

Address: ____

Phone: _______________ Email: ____

Price: ___________________________ Deposit Amount: ______________________

Balance: _________________________ Date Paid: ____________________________

Hours Worked: ____

Description of Installation: ____

____

____

____

Materials Installed: ____

____

____

____

____

Colors/Finishes: ____

____

____

____

Notes: ____

____

____

____

Installation

Start Date: ______________________ Completion Date: ______________________

Customer Name: ____

Address: ____

Phone: _______________ Email: _______________________________________

Price: __________________________ Deposit Amount: _____________________

Balance: _________________________Date Paid: _________________________

Hours Worked: ____

Description of Installation: ____

____

____

____

Materials Installed: ____

____

____

____

____

Colors/Finishes: ____

____

____

____

Notes: ____

____

____

____

Index

A

Customer Name: ______________________________________Page: __________

Customer Name: ______________________________________Page: __________

Customer Name: ______________________________________Page: __________

Customer Name: ______________________________________Page: __________

Customer Name: ______________________________________Page: __________

Customer Name: ______________________________________Page: __________

Customer Name: ______________________________________Page: __________

Customer Name: ______________________________________Page: __________

Customer Name: ______________________________________Page: __________

Customer Name: ______________________________________Page: __________

B

Customer Name: ______________________________________Page: __________

Customer Name: ______________________________________Page: __________

Customer Name: ______________________________________Page: __________

Customer Name: ______________________________________Page: __________

Customer Name: ______________________________________Page: __________

Customer Name: ______________________________________Page: __________

Customer Name: ______________________________________Page: __________

Customer Name: ______________________________________Page: __________

Customer Name: ______________________________________Page: __________

Customer Name: ______________________________________Page: __________

Index

C

D

Index

E

Customer Name: ______________________________ Page: __________

Customer Name: ______________________________ Page: __________

Customer Name: ______________________________ Page: __________

Customer Name: ______________________________ Page: __________

Customer Name: ______________________________ Page: __________

Customer Name: ______________________________ Page: __________

Customer Name: ______________________________ Page: __________

Customer Name: ______________________________ Page: __________

Customer Name: ______________________________ Page: __________

Customer Name: ______________________________ Page: __________

F

Customer Name: ______________________________ Page: __________

Customer Name: ______________________________ Page: __________

Customer Name: ______________________________ Page: __________

Customer Name: ______________________________ Page: __________

Customer Name: ______________________________ Page: __________

Customer Name: ______________________________ Page: __________

Customer Name: ______________________________ Page: __________

Customer Name: ______________________________ Page: __________

Customer Name: ______________________________ Page: __________

Customer Name: ______________________________ Page: __________

Index

G

H

Index

I

Customer Name: ______________________________Page: __________

Customer Name: ______________________________Page: __________

Customer Name: ______________________________Page: __________

Customer Name: ______________________________Page: __________

Customer Name: ______________________________Page: __________

Customer Name: ______________________________Page: __________

Customer Name: ______________________________Page: __________

Customer Name: ______________________________Page: __________

Customer Name: ______________________________Page: __________

Customer Name: ______________________________Page: __________

J

Customer Name: ______________________________Page: __________

Customer Name: ______________________________Page: __________

Customer Name: ______________________________Page: __________

Customer Name: ______________________________Page: __________

Customer Name: ______________________________Page: __________

Customer Name: ______________________________Page: __________

Customer Name: ______________________________Page: __________

Customer Name: ______________________________Page: __________

Customer Name: ______________________________Page: __________

Customer Name: ______________________________Page: __________

Index

K

Customer Name: ______________________________________ Page: __________

Customer Name: ______________________________________ Page: __________

Customer Name: ______________________________________ Page: __________

Customer Name: ______________________________________ Page: __________

Customer Name: ______________________________________ Page: __________

Customer Name: ______________________________________ Page: __________

Customer Name: ______________________________________ Page: __________

Customer Name: ______________________________________ Page: __________

Customer Name: ______________________________________ Page: __________

Customer Name: ______________________________________ Page: __________

L

Customer Name: ______________________________________ Page: __________

Customer Name: ______________________________________ Page: __________

Customer Name: ______________________________________ Page: __________

Customer Name: ______________________________________ Page: __________

Customer Name: ______________________________________ Page: __________

Customer Name: ______________________________________ Page: __________

Customer Name: ______________________________________ Page: __________

Customer Name: ______________________________________ Page: __________

Customer Name: ______________________________________ Page: __________

Customer Name: ______________________________________ Page: __________

Index

M

N

Index

O

P

Index

Q

Customer Name: ______________________________ Page: __________

Customer Name: ______________________________ Page: __________

Customer Name: ______________________________ Page: __________

Customer Name: ______________________________ Page: __________

Customer Name: ______________________________ Page: __________

Customer Name: ______________________________ Page: __________

Customer Name: ______________________________ Page: __________

Customer Name: ______________________________ Page: __________

Customer Name: ______________________________ Page: __________

Customer Name: ______________________________ Page: __________

R

Customer Name: ______________________________ Page: __________

Customer Name: ______________________________ Page: __________

Customer Name: ______________________________ Page: __________

Customer Name: ______________________________ Page: __________

Customer Name: ______________________________ Page: __________

Customer Name: ______________________________ Page: __________

Customer Name: ______________________________ Page: __________

Customer Name: ______________________________ Page: __________

Customer Name: ______________________________ Page: __________

Customer Name: ______________________________ Page: __________

Index

S

T

Index

U

V

Index

W

Customer Name: ____________________________________ Page: __________

Customer Name: ____________________________________ Page: __________

Customer Name: ____________________________________ Page: __________

Customer Name: ____________________________________ Page: __________

Customer Name: ____________________________________ Page: __________

Customer Name: ____________________________________ Page: __________

Customer Name: ____________________________________ Page: __________

Customer Name: ____________________________________ Page: __________

Customer Name: ____________________________________ Page: __________

Customer Name: ____________________________________ Page: __________

X

Customer Name: ____________________________________ Page: __________

Customer Name: ____________________________________ Page: __________

Customer Name: ____________________________________ Page: __________

Customer Name: ____________________________________ Page: __________

Customer Name: ____________________________________ Page: __________

Customer Name: ____________________________________ Page: __________

Customer Name: ____________________________________ Page: __________

Customer Name: ____________________________________ Page: __________

Customer Name: ____________________________________ Page: __________

Customer Name: ____________________________________ Page: __________

Index

Y

Z

Made in the USA
Columbia, SC
09 December 2022